Nature's Children

LIZARDS

Bill Ivy

Grolier

FACTS IN BRIEF

Classification of lizards

Class:	*Reptilia* (reptiles)
Order:	*Squamata* (reptiles with scales)
Suborder:	*Sauria* (lizards)
Family:	There are 20 families of lizards.
Genus:	There are 41 genera of lizards.
Species:	There are over 3000 species of lizards.

World distribution. Lizards are found throughout the world, except in polar regions.

Habitat. Varies with species.

Distinctive physical characteristics. Most have four legs, a long tail and external ear openings and eyelids; body is covered with dry scales.

Habits. Vary with species.

Diet. Varies with species.

Published originally as
"Getting to Know . . . Nature's Children."

This series is approved and recommended by the Federation of Ontario Naturalists.

This library reinforced edition is available exclusively from:

Grolier Educational Corporation
Sherman Turnpike, Danbury, Connecticut 06816

Printed and Bound in U.S.A. ISBN 0-7172-2638-7

Contents

If you have ever been lucky enough to see a lizard in the wild, you probably just got a glimpse of it as it scurried by. Lizards are both shy and extremely quick, and most people know very little about them. This is unfortunate since lizards are among the most amazing animals in the world. They come in all sizes, shapes and colors. Some are tiny, legless, worm-like creatures while others are the size of a large dog.

There are lizards that can sail through the air, deep sea dive, change color, walk on ceilings or eat pigs. To find out more surprising things about lizards, read on.

Land of the Lizard

Lizards are found in all parts of the world except in polar regions. In fact, there is at least one species of lizard on every continent except Antarctica. About 3000 species have been identified so far.

The majority of lizards live in tropical and sub-tropical climates, where it is warm all year round. Some can also be found, however, in areas where winters are cold. There they must hibernate underground during the winter months to survive.

Wherever they live, these incredible creatures make themselves at home in a wide variety of environments, from tropical rain forests to deserts and from mountains to lowland plains.

In North America, lizards live as far north as southern Canada.

This anole blends in well with its chosen resting spot.

Lizard Relatives

Lizards are reptiles, as are snakes, crocodiles, alligators and turtles. Like all reptiles, lizards are cold-blooded, which means they cannot generate their own body heat. But lizards are not usually cold—they get warmth from their surroundings. Lizards lie in the sun to warm up and rest in the shade to cool down. Some types can change color to regulate their temperature—they turn darker to absorb more heat or lighter to absorb less.

Lizards are often mistaken for salamanders, but there are several ways to tell them apart. Salamanders, like frogs, are amphibians. They have smooth, moist skin, while lizards are covered with scales. Salamanders have four fingers on their front limbs, lizards have five. Also, while lizards love being in the sun, salamanders avoid it.

Although some lizards are legless and look like snakes, there are a few ways to tell who's who. Snakes do not have moveable eyelids, but most lizards do and most lizards have ear openings whereas snakes do not.

Opposite page: *The red-spotted newt (bottom), like most salamanders, spends part of its life underwater, breathing through gills. Lizards never have gills. Few of them spend any time at all in water, and those that do have to come up regularly for air. (Top: common iguana)*

Scaly Skin

Have you ever held a lizard? If not, what do you think it feels like? Wet and slimy? Not at all. Lizards are covered from head to tail with scales and are dry to the touch.

Scales are made of keratin—the same thing your fingernails are made of. The scales of some lizards are smooth and shiny, while those of other types are as rough as a pine cone. No matter what they look like, scales give the lizard a thick waterproof coat. And they help keep moisture in so that the lizard doesn't dry out in the sun.

What do you think happens as a lizard grows bigger and its scaly skin becomes too small? It sheds the outer layer for a new, larger coat that has been forming underneath.

The chameleon, like most lizards, sheds its skin in patches rather than all at once the way snakes do.

Sense Abilities

Most lizards have excellent vision, and by looking at their eyes you can often learn a lot about their habits. Lizards with round pupils are commonly active during the day, while those with vertical slits are usually creatures of the night. Burrowing lizards have transparent eyelids that protect their eyes from sand and dirt.

Lizards also have a sharp sense of smell. Like you, they smell with their nose, but some also pick up scents in another way—by sticking out their tongue. When a lizard flicks out its tongue, it picks up tiny scent particles from the air. In the lizard's mouth, two special organs called Jacobson's organs allow the lizard to taste the air. Lizards may use their tongue to smell nearby food or to sense what kind of environment they're in, but only one type, the chameleon, uses its tongue to actually catch its food.

Lizards have ears but they don't have ear flaps as you do. Their ear openings are on the same level as their skin or just below it. Burrowing lizards have small ear openings or none at all.

Opposite page:
If you look closely you can see the ear opening on this emerald swift's head at the edge of its mouth.

Opposite page:
If you're a hungry iguana, nothing hits the spot like a crunchy katydid.

On the Menu

How would you like this for dinner—ants, centipedes, spiders, worms and slugs? Not very appetizing, you say? To a lizard this would be a feast!

Many lizards feed almost entirely on insects and other small invertebrates. Some prefer a diet of plants or fruit, however. Marine iguanas, for instance, eat only algae. The larger lizards also eat animals such as clams, mice and even other lizards. One species, the Komodo dragon, eats dead animals and occasionally catches a small wild pig or a domestic sheep!

Lizards do not have very good table manners and often just swallow their food whole. Large prey is pulled apart and eaten a mouthful at a time.

Most lizards get the water they need by lapping up raindrops and dew. But some desert species rarely get a chance to drink. Fortunately, the plants they eat usually provide enough water for their needs.

Happy Birthday

During the mating season, a male lizard chases a female and they mate. The couple may stay together for one or more seasons.

Lizards, like birds, produce eggs. The eggs are usually oval shaped with shells that are soft and leathery. A female may lay from 1 to 25 eggs, depositing them most often in a warm humid place, such as beneath leaf litter and logs.

A few species of lizards guard their eggs, but most just leave them to hatch on their own. When the baby lizard is ready, it uses the razor-sharp egg tooth on its snout to break free of its shell. The egg tooth falls off within a few days. Newborn lizards, most of which look a lot like their parents, are able to take care of themselves as soon as they hatch. They will mature three months to two years later.

Some species of lizards, especially in cold climates, keep the eggs in their body until they hatch, then give birth to live young. They do this so that the eggs stay warm.

Opposite page: *A common iguana emerges from its shell looking like a small version of its parents. On the other hand, young Jackson's chameleons (overleaf) obviously change a good deal as they mature.*

When faced with an enemy, Australia's frilled lizard extends its frill and opens its mouth wide to make itself look as big and as menacing as possible.

Self-Defense

If you were a lizard you would not want to meet a raccoon, otter, fox, opossum, snake, hawk or badger. Any one of these animals would enjoy having you for lunch. So might a few lizards that are bigger than you.

Fortunately, lizards have a few tricks to help them survive these dangers. When it senses an enemy, a lizard may freeze. Because its color often blends in with its surroundings, it can be very hard to see when it is not moving. If this doesn't work, the lizard may run, hide, hiss, play dead or even squirt blood from its eyelids.

Each species has its own way of protecting itself. But the most amazing defense of all is common to many species—a detachable tail! When grabbed from behind the lizard's tail breaks off cleanly and wiggles around on the ground by itself. Usually this distracts the lizard's enemy long enough for the lizard to escape. This in no way harms the lizard, as a brand new, shorter tail will grow back within a few weeks.

"True" Lizards

When most people think of lizards they probably picture one with four legs and a tail that is usually longer than its body. This is the classic description of the family *Lacertidae*, also sometimes called "true" lizards, probably because they were some of the first to be studied by scientists. They are small- to medium-sized creatures and many are brilliantly colored.

Lacertids eat insects, spiders, scorpions and small vertebrates. They are found in Europe, Asia and Africa, and can live as far north as the Arctic Circle.

This lacertid is known as the green lizard. It's easy to see why!

Most geckos are brown or gray and active mainly at night, but this Mauritius day gecko is brightly colored and prefers to be up and around in the daytime.

House Guests

While most lizards try to stay as far away from people as possible, geckos often move right in with them. In hot countries where they thrive, "house geckos" may even live like pets in people's homes. They earn their keep by eating insects. Thanks to special pads on the bottom of their feet they can walk straight up walls or across ceilings, just like a fly. And they have retractable claws for rough surfaces such as trees.

Most geckos are between 5 and 15 centimetres (2 to 6 inches) long. Their skin is almost transparent and only a few have moveable eyelids. The others have clear scales called spectacles that cover and protect their eyes.

Geckos are the most vocal of all the lizards. The sounds they make range from the hoarse "geh-ho" that gives them their name to soft chirps and loud shrieks.

Quick Change Artists

True chameleons live in Africa and Asia and are famous for their ability to change color. Although they may blend in with their surroundings, this is not the reason for the color change. They are simply reacting to changes in temperature, light conditions or even moods.

Most chameleons live in trees and can grip branches with their toes and tail. They can also look in two directions at the same time, and many have horns or spines on their head. Unlike many lizards, however, chameleons cannot grow a new tail if they lose the one they have.

Chameleons eat insects but they move so slowly it's a wonder they ever catch anything. If it weren't for their amazing tongue they probably wouldn't. Remaining perfectly still, the chameleon waits for an insect to come within striking range. In a flash it shoots out its tongue—which is at least as long as its body—and captures the insect with the sticky tip. It reels in the catch and dinner is served.

Zap!

Slinky Skinks

The largest family of lizards in the world is the skink family. There are over 1200 different species of skinks, and they are found on every continent except Antarctica. Most make their home in climates that are warm year-round, but a few live in southern Canada where winters are cold. These species hibernate all winter.

Skinks are sleek creatures with rounded bodies. Most are quite small and have four small legs and a tail that is longer than their body. Burrowing skinks have tiny legs or no legs at all, however, and they have a clear window in their lower eyelid so that they can see underground when their eyes are closed. Some skinks eat only plants, while others eat insects, spiders, worms and even small mice.

Although they are quite active, skinks are secretive creatures and so are rarely seen. If you want to see one for yourself, try looking under rocks, in moist decaying logs or under leaves. If you do find one you'd better not grab it. Not only will it drop its tail—it may bite.

Opposite page:
While most skinks are slim and have long tails, the stump-tailed skink is an exception. As its name suggests, it has a short, stumpy tail, which in fact looks a lot like its head. Can you tell which end of these skinks is which?

Mighty Monitors

Meet the giants of the lizard world, the monitors. Although most are not much more than 30 centimetres (a foot) long, some of these beasts can grow to a length of 3 metres (10 feet)!

The largest monitor, the Komodo dragon, has been known to eat a sheep for dinner, and all of them will eat just about anything they can catch. A special lower jaw which can unhinge from the skull and swing down enables them to swallow much of their prey whole and the largest in a few big chunks.

There are about 30 species of monitor lizard, all of them living in the tropics. They can be found in fields or jungles, and many spend a lot of time in trees. Most are good swimmers.

A Komodo dragon may weigh up to 135 kg (300 lb)!

Whiptail.

Fleet of Foot

When it comes to speed not many reptiles can match the members of the teiid family. With their long, slender bodies, straight legs and extra long tails, these lizards are built for speed. Some species can reach up to 29 kilometres (18 miles) per hour. That's an incredible 8 metres (27 feet) a second. Bet you can't move that fast!

These active lizards range in size from 10 to 120 centimetres (4 to 48 inches), but most come about halfway in between. Because of their quickness and long skinny tail some are known as racerunners and whiptails. The scales on their upper body are small and rounded and don't overlap. This allows them to move easily and gives them a velvety appearance.

Most whiptails eat insects, worms and snails, although larger teiids eat small mammals, birds, eggs and other reptiles. In all, there are over 200 members of this family. Most are found in South America, with a few in the United States.

Opposite page: *The common iguana is also known as the green iguana—for obvious reasons.*

Iguanas, Big and Little

The iguanids are a large family of lizards—there are almost 700 species. Most are found in North and South America. The smallest ones are a mere 10 centimetres (4 inches) in length, and the largest are 2 metres (6 feet) long. While most eat plants and fruit, others, especially young iguanas, eat insects and other small invertebrates.

The most impressive member of this family is the common iguana, the largest lizard found in the Americas. It looks like a dinosaur with a fringe of scales running down its back to the base of its tail. Like most iguanas, it has a dewlap, or loose flap of skin, under its throat. This dewlap can be puffed out when the lizard wants to challenge another lizard or impress its mate.

With its sharp claws the common iguana is an excellent climber and can scamper up trees like a squirrel. Despite its size it is very quick and can outrun a dog on the ground. Not bad for a lizard!

Deep Sea Divers

You might not expect to find lizards in the ocean, but that is where marine iguanas spend a lot of their time. They live on the rocky beaches of the Galapagos Islands in the Pacific Ocean.

A marine iguana is perfectly suited for its life in and around the water. It has a long, flat tail that it uses to push itself through the water and long curved claws that help it climb out onto the rocks.

Marine iguanas feed on seaweed and algae. Although they prefer to eat along the shoreline, at times they must dive as deep as 16 metres (50 feet) to find food. These incredible divers can stay underwater for almost an hour before surfacing for air. They live in large groups called colonies. In a typical colony the rocks are covered with wall-to-wall lizards.

No, it's not a dinosaur—it's a marine iguana.

Worm Lizards and Horned Toads

Although most lizards live on the ground or in trees, some types actually spend their whole life underground. People often mistake these burrowing lizards for worms. For this reason, they are known as worm lizards. Their bodies have adapted to life below ground—most have no legs, which makes it easier for them to move through soil. Their eyes are very small because they are rarely used, and many have pale skin. They may be up to 40 centimetres (16 inches) long and they feed on small insects.

The horned lizard is another very unusual type. Although it is also called a horned toad, it is not a toad but a lizard with a short tail. It is 7.5 to 18 centimetres (3 to 7 inches) long and has a flat body with sharp spikes on its head and back to protect it from enemies. But that's not its only method of self-defense. Some horned lizards can also squirt blood from their lower eyelids. Many animals, such as wild dogs, lose their appetite for lizard after such a display.

Horned lizard.

Leaping Lizards

You know that lizards can swim, run, burrow and climb trees, but did you know that one species can glide through the air? Known as the draco, or flying dragon, it is a member of the agamid family and lives in Asia and Indonesia.

The draco has long, flexible ribs that can be spread out from its body like an umbrella, revealing a large pair of skin flaps. Leaping from a branch high in a tree, it glides through the air like a paper airplane. When at rest the draco is rather dull-colored and looks more like an insect than a lizard.

It's a bird . . . it's a plane . . . it's a flying draco!

Handle with Care

The anguids are a very old family of lizards. They have tiny legs and toes, or they may have no legs or feet at all. Tough scales form rings around their body like a coat of armor.

One of the most interesting anguids is the glass lizard. Also known as the glass snake, this bizarre creature looks like a snake and has shiny, smooth scales. It gets its name from the fact that its fragile tail will break off if handled roughly. Its flickering tongue makes it appear even more snake-like. Glass snakes range from 60 to 90 centimetres (2 to 3 feet) in length and eat insects, small mammals and other lizards.

Would you recognize this glass snake as a lizard? Well, that's what it is.

It's best to keep your distance from the Gila monster.

Lethal Lizards

While most lizards are harmless, there are two species you wouldn't want to mess around with. They are the Gila (pronounced hee-luh) monster and the Mexican beaded lizard, and they're the only poisonous lizards in the world. Both species live in the desert and are quite colorful. They are mainly nocturnal, sleeping most of the day away in a moist, dark burrow.

The 60-centimetre (2-foot) long Gila and its slightly larger Mexican cousin have cheeks that appear to be swollen. This is where their venom glands are. Unlike snakes, these lizards do not have long fangs for injecting poison into their victims. Instead, they have special grooved teeth. When they bite they hang on for as long as possible while the venom flows from their mouth into the wound. Although the poison can kill small animals it is rarely fatal to people.

Helping Out

Now that you know more about lizards you would probably like to meet one face to face some day. And if you are alert enough and lucky enough to do so, you are almost sure to enjoy the experience since lizards are not only fascinating but can also be quite beautiful.

However, if you are still a little nervous about their scaly skin, unusual habits and sharp claws, remember that almost any lizard will be at least as nervous about you. Remember too that lizards are very useful. They eat millions of insects and other small pests that may ruin crops or be harmful to other animals and to people.

In some countries, lizards are even thought to bring good luck. So much so, in fact, that in parts of Southeast Asia people are reluctant to move into a new house until a tokay, a type of gecko, has moved in. What a strange and wonderful house guest to have!

Words to Know

Amphibians A group of animals that live both on land and in water. Frogs, toads and salamanders are amphibians.

Cold-blooded Term used for animals that have no automatic internal control of their body temperature.

Dewlap A flap of loose skin that hangs from the neck of some animals.

Gland A body organ where certain substances are made.

Hibernate To go into a deep sleep for the entire winter.

Invertebrate An animal without a backbone.

Keratin Protein that forms the basis of horns, nails, hair, feathers and scales.

Mate To come together to produce young. Either member of an animal pair is also the other's mate.

Prey Animal that other animals hunt for food.

Reptile Class of cold-blooded animals that includes snakes, alligators, turtles and lizards.

Salamander An amphibian with a tail that has moist, scaleless skin and breathes through gills in the larva stage.

Scale Thin, hard overlapping plate that protects a lizard's skin.

Venom Poisonous fluid produced by some lizards.

INDEX

Cover Photo: Bill Ivy

Photo Credits: Boyd Norton, pages 4, 31; Bill Ivy, pages 7, 8a, 8b, 11, 12, 28, 39, 44; Doug Wechsler (Vireo), page 15; K.H. Switak (Photo Researchers), page 16; Zoological Society of San Diego, pages 18-19, 35, 42; Stock Market Inc., page 20; Fco. Márquez, page 23; Tom McHugh (Photo Researchers), pages 24, 41; Stephen Dalton (Photo Researchers), page 27; Ron Garrison (Zoological Society of San Diego), page 32; Ron Dengler (Network), page 36.